Build a Clock
Ideas Book

written by

Toni and Mark Tippett

Published by Hands On

© 2000 Hands On, Tonbridge, Kent
All rights reserved. No part of this publication may be reproduced without the express permission of Hands On unless the page is labelled photocopiable.

Special thanks to Richard Shestopal of Shesto Ltd., Sally Isaac, Karen Walsh, C.R. Clarke, Terry Hardy Photography, Roger Brandon and Chedworth Roman Villa, © The National Trust

Hands On is a registered Trademark and is part of the Commotion Group,
Unit 11 Tannery Road, Tonbridge, Kent TN9 1RF

ISBN: 1-873101-30-9

CONTENTS

	Introduction 5
	Basics 6
NATURE	 9
	Wavy Clock 10
	Hand Made Paper Clock 12
	Sun and Moon Clock 14
	Art Roc® Clock 16
TEXTURES	 19
	Textured Clock 20
	Air Drying Clay Clock 22
	Folded Clock with Etched Finish ... 24
	Mosaic Clock 26
ARCHITECTURE	 29
	Traditional Mantle Clock 30
	Mantle/Desk Clock 32
	Arched Metal Clock 34
	Cut and Bend Clock 36
	Acrylic Clock 38
	Copper Clock 40
ANIMALS	 43
	Fun Pendulum Clocks 44
	Whale Clock 46
	Crocodile Clock 48
NOVELTY CLOCKS	 51
	ZZZzzzz Clock 52
	Say Cheese! 54
	Big Numbers Clock 55
	Making Clocks Hints & Tips 57
	Clock Face Numerals Template 60

INTRODUCTION

If You Can Drill a Hole You Can Make a Clock…

Designing and then building a working clock is a popular and surprisingly easy project that can be undertaken by anyone who can drill a hole! Precision quartz mechanisms are readily available and will keep accurate time. They can be purchased with the fixing nuts, basic hands, hanger etc. and only require an AA battery to begin to tick.

Anything can be used to make a clock face. It is simply a matter of letting your imagination run wild to come up with something novel, aesthetic or even comical! The material could be paper or card, a piece of metal, plastic, wood or conceivably, even concrete! The design could be inspired by your favourite cartoon character, a pet, the shapes found in nature, a texture, any shape with a strong profile ..... literally anything.

To fix the mechanism into the clock face requires a hole 8mm in diameter to allow for the spindle to be threaded through. It is fastened into place using a front connecting nut which holds the mechanism tight against the back of the clock face. The hour, minute and second hands are then push-fitted onto the spindle.

This book contains over twenty clock designs which have been inspired by a variety of shapes including nature, architecture, animals etc. As a teacher of design, I have used my experience of various materials to show how to make decorative clock faces including various types of paper, card, wood, plaster, plastic and metal. I have applied a range of techniques and finishes to give different effects to act as an index of possible outcomes to inspire someone new to designing and making. The clocks in this book would make stunning features in any home or office either for yourself or as a gift.

We hope you find this book interesting and inspiring when clock making. I have listed some suppliers at the back of the book who will be pleased to help you with the mechanisms, hands and materials mentioned.

We wish you every success.

Toni and Mark Tippett

BASICS

The Clock Mechanism

All of the clocks in this book use precision quartz mechanisms. These are readily available and surprisingly low cost. They are powered by an AA battery. You can also buy these movements with integral pendulums or separate pendulums which can sit away from the clock face (see below). Basic hands are often supplied with the mechanism, but a large range of elaborate hands are also available to suit the style of clock you choose to make (see 'Clock Hands' opposite).

When fitting the mechanism into your clock it is always best to follow the instructions provided with it as they may vary slightly.

Choosing the Spindle

The spindle is the shaft that sticks out of the middle of the movement, and to which the hands are fixed. As the spindle has to go through the dial (face) of the clock and stick out enough on the other side for the hands to be fitted onto it, the length of the spindle is very important. A spindle length of 16mm will take a clock face up to 8.6mm thick, and a spindle of 21.5m in length a clock face of up to 12.5mm thick. If you are making a clock from a very thin material such as card, you should choose a mechanism with a shorter spindle length.
The diameter of the hole required is 8mm to allow for the spindle to be threaded through.

Varying spindle lengths: 12mm, 16.5mm, 21.5mm, 26.5mm

Pendulums

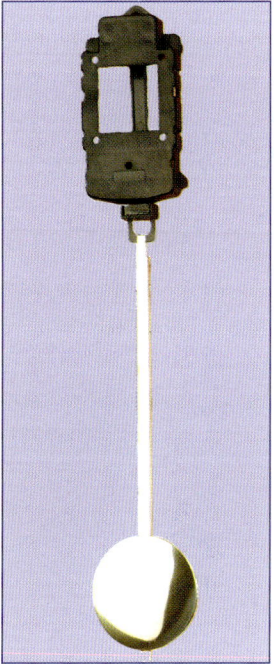

By attaching a self-driven independent pendulum movement, you can add a swinging feature. Most people have seen brass pendulum 'bobs' on traditional clocks, but let your imagination run wild and use the swinging action for a whole variety of other animated uses – wagging tails of dogs, cats, mice or a foot kicking a football. Pendulum drive units can be placed around a standard movement – or they can be used on their own. They work completely independently of the clock movement with a separate battery, so they can be mounted anywhere.

Four of this 'scary spiders' legs swing using an integral clock/pendulum movement

Independent clock pendulum

BASICS

Clock Hands

Hands should be chosen to suit the style of clock you are making and to fit the diameter of the face. Simple straight (baton) hands are best suited to contemporary looking clocks and fancy hands look better on ornate style clocks. Ensure you make the clock face big enough so that the hands do not protrude over the edge which will not look right and increases the possibility of damage. Many clock hands can be cut with scissors to shorten them or painted and decorated to become a feature.

Care must be taken when fitting the hands to the mechanism – follow the instructions supplied. Too much force applied when fitting the hands could prevent the mechanism from working properly.

Numerals

For many of the clocks in this book I have used my own markers, shapes and numerals for the clock faces, and in many cases, the design looks better without any markings. To give your clock a very professional looking finish, you can buy self-adhesive Arabic and Roman numerals which are quick and easy to use.

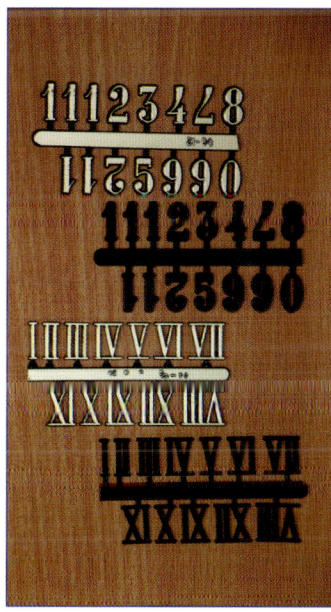

Marking out the Number Settings

The easiest and simplest way to mark out the number spacings on a clock face is to use a mathematical 360° angle measure or a protractor. Place the angle measure (which is clear plastic) over the clock face, lining up both centre points and mark off every 30° to give you the correct numeral spacing. (You could also use a template such as the one found at the back of this book).

WAVY CLOCK

Clock Project **1**

Starting Points — I have used the sea and beach as my theme for this clock. It has been influenced by the shape of the sea waves, and the washed up driftwood you can find on a beach, but you don't need to live close to a beach to make this clock!

Design Features — This clock is made using undressed timber which has been given a white washed finish to achieve the bleached effect, producing an understated elegant design. The shape of the clock has been influenced by sea waves. It uses an insert clock mechanism which has been slotted into the timber.

WAVY CLOCK

Tools and Equipment

Drill and drill bit
Glasspaper
Jigsaw
Paint brush
Pencil
Safety goggles

Materials

Beeswax (optional)
Clock face
Undressed timber
White or cream emulsion

Equipment and materials used to make this clock.

Instructions For Making

1. Mark out the design onto the timber.
2. Cut out the design.
3. Drill a hole through the area where the clock face will be inserted, big enough to get the jigsaw blade through.
4. Cut out a circle the right size for your clock face.
5. Smooth the timber lightly to take the edge off the roughness when touched, and round the edges to soften the lines.
6. Dilute the emulsion (one part paint to two parts water) and paint the timber all over. You can repeat this process until you get the depth of colour wash you require.
7. Apply a very fine coat of beeswax over the painted timber.
8. Slot the clock face into position.

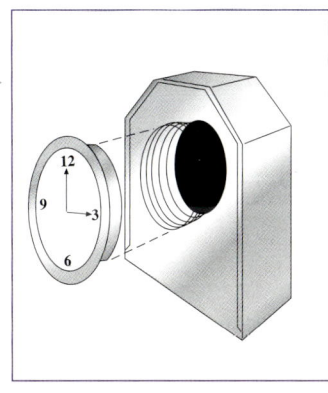

 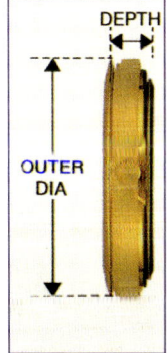 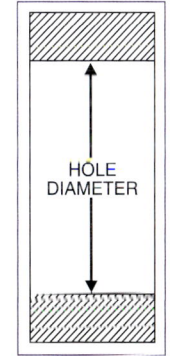 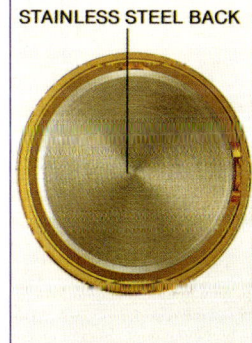

Cut away showing hole through material

 **TIPS**

If you go to a small timber merchants you can buy just the amount of timber you need rather than a whole length.

Further Things to Try

You could use real driftwood to make this clock, in which case you may not want to colour wash it, but rather leave it as it is or simply give it a fine coat of beeswax to enhance it.

HAND MADE PAPER CLOCK

Clock Project **2**

Starting Points	Seasons change with time, and with their arrival they bring their own colours and textures. I especially like autumn with its warm, rich tones of rusty brown and yellow, as the leaves fall off the trees. It is this aspect of autumn which I used in my design for this clock. You can choose a season and depict a simple motif, like the leaves I have used, and combine it with the colours of that season to make your clock.
Design Features	I used handmade paper to make this clock because it has a natural and light appearance. Although the shape of the design is bold, it looks graceful and has a delicate texture. It will particularly appeal to those who appreciate the beauty of handmade papers and wish to experiment with making their own. The range of textures and colours you can achieve when making paper is enormous. At the back of this book I have described how to make the paper I used in my clock design. This basic paper making process can be as complex and varied as required. There are a range of books available on paper making which will give you endless ideas for further experiments.

HAND MADE PAPER CLOCK

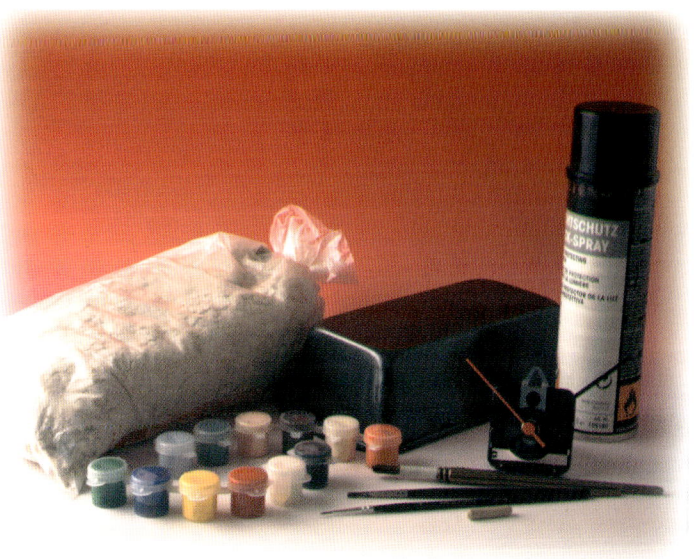

Equipment and materials used to make this clock.

Tools and Equipment

Drill and drill bits
Paper making equipment
Paint brushes
Wooden mould

Materials

Clock mechanism
Clear spray varnish
Paper pulp/various materials
Paints of your choice (optional)
Any cheap/scrap timber for the mould
15mm length of dowel the diameter of the spindle on your clock mechanism

Instructions For Making

1. Make your wooden mould using the timber. The depth of the mould should be the same as the depth of your clock mechanism to enable it to hang on the wall. Drill and fix the dowel in the place that will be the centre of your clock face. (This will form the hole for the clock mechanism).
2. Make some paper pulp (refer to page 59).
3. Cover the wooden mould with the paper pulp, carefully pressing the pulp firmly together, and ensuring no gaps are left.
4. Leave the paper to dry on the mould.
5. When the paper is dry, carefully remove it from the mould.
6. At this stage you can add any paint detail you wish.
7. To protect the paper from getting marked and to give it extra strength, spray the whole clock with a clear lacquer, and leave it to dry.
8. Fit the clock mechanism and hands into place.

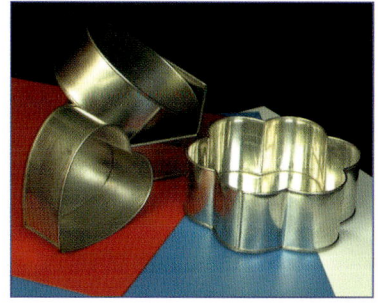

Use a baking tray as a mould

Painting detail onto paper

  **TIPS**

This is an excellent way to recycle paper.

Further Things to Try

You could change the wooden mould to any shape you like, or you could try using an object you may already have at home such as a baking tin as a mould. This design will allow you to try out a wide variety of different handmade papers, which will give the clock a different look, colour and texture.

SUN AND MOON CLOCK

Clock Project 3

Starting Points | The sun and moon define our nights and days. They have been used to measure time for millions of years. I took these two elements of time-keeping and drew each of them onto a piece of paper, both the same size. I then cut them out and overlapped them to form this design. This is a simple technique you can use to create your own designs. You can use more than two pictures if you wish. You could also use pictures of different sizes.

Design Features | The clock is made from one piece of MDF (Medium Density Fibreboard) which has been cut to give the clock face a carved effect. The MDF is very easy to work in this way, making this design suitable for those with limited woodworking skills.

SUN AND MOON CLOCK

Tools and Equipment

Pencil
Ruler
Compass
Drill and drill bit
Jigsaw or coping saw
Glasspaper
Craft knife or chisel

Materials

Clear spray varnish
Clock mechanism
MDF
Undercoat paint (white emulsion will provide a base coat)
Watercolour paints and brushes

Equipment and materials used to make this clock.

Instructions For Making

1. Mark out the design onto the MDF.
2. Cut out the outer profile of the shape.
3. Using a craft knife, cut along the line between the sun and the moon. Then carefully start to lift away layers of the MDF from the sun by easing the blade into the edge of the MDF. Because of the way the MDF is formed, it will peel away in layers. To cut deeper, keep scoring the line between the sun and moon to enable you to take further layers away. Try to get an uneven surface over the entire sun, you should be left with a slightly 'furry' texture. Take extra care when using a cutting knife and always cut with the blade facing away from you.
4. Smooth the edges of the clock, but not the front or back surfaces.
5. Drill the hole for the mechanism.
6. Give the entire clock a coat of undercoat.
7. Paint the clock. I used watercolours which give a subtle, more translucent effect
8. When dry, finish by spraying with clear varnish.
9. Fit the clock mechanism into place.

Stages of Cutting Relief Layers from MDF

Draw your design onto the MDF

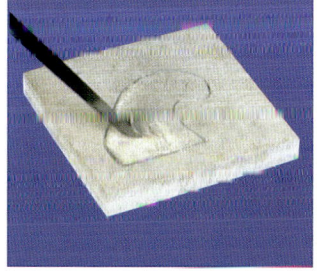
Carefully peel off the layers of MDF using a chisel or cutting knife

Showing layers peeled away.

TIPS

Always cut away from you and clamp your work securely. Do not use a 'snap-off' blade to pick away at the surface of MDF.

Further Things to Try

You can use this method of working MDF on any design you want. It is a very easy way to give the wood a relief surface without the need for any specialist tools or woodworking skills.

'ART ROC' CLOCK

Clock Project 4

Starting Points	'Art Roc®' or 'Modroc®' comes as a bandage which is plaster impregnated. To use it, the material is simply dipped into water and then draped over a former. When dry, the plaster sets solid into the shape of the former. The materials could be placed over a wooden block, a baking tray (to be removed later) or, better still, the former made from wire or chicken wire, which remains forever cast inside the material once set.
	This is a good material for forming over a mould to achieve three dimensional forms. It is easy to work with and dries very fast. It is another material that is easy for everyone to use, even children will have great fun using it. The two designs here are based on the sky, sunsets, stars and clouds.
Design Features	I used an integral clock/pendulum mechanism for the design shown above, which was painted using emulsion and gold paint (tester size pots are very economical as is any left over paint). I then added self adhesive gold stars and Roman numerals. The clock shown opposite I painted with acrylic paints (which have a sheen) and used self-adhesive numbers for the clock face.

'ART ROC' CLOCK

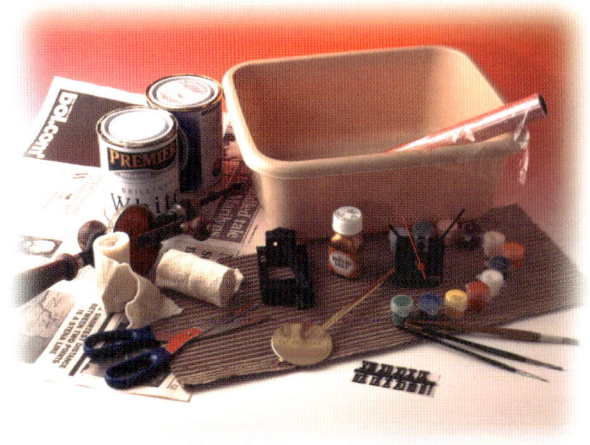

Equipment and materials used to make this clock.

Tools and Equipment

Bowl of cold water
Cling film
Hand drill or craft knife
Old newspapers
Scissors

Materials

Art Roc
Clock mechanism
To make a mould: used corrugated cardboard held into shape with masking tape, or use an existing object such as a bowl.
Paints and brushes of your choice
Self adhesive numbers or numerals

Instructions For Making

1. Make your mould to the size and shape you want your clock to be. I modelled my mould using pieces of corrugated cardboard taped together with masking tape. If you need a more versatile material to get curves for example, then clay or chicken wire will work very well.

2. I cover the my mould with cling film so it is easier to remove when it is dry.

3. Cover your work surface with the old newspapers. Follow the instructions that come with your Art Roc. I cut strips to the size I required before I started. I was then able to dip each one into the cold water and apply it over my mould. I built it up with a few layers to give it strength, and then left it to dry whilst still on the mould.

4. When it is dry and hard, remove the clock from the mould, easing it gently away using the cling film.

5. Make the hole for the mechanism carefully using a hand drill or craft knife.

6. Paint the clock.

7. Fit the clock mechanism into place.

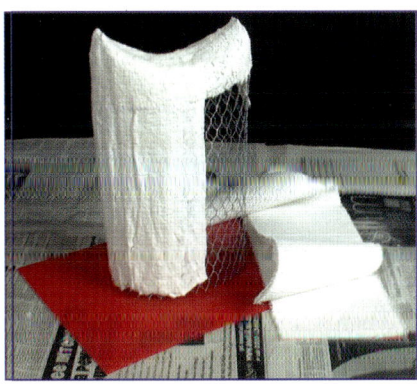

Applying 'Art Roc' over chicken wire mould

TIPS

Put lots of newspaper out when you make this clock as it can get very messy!
Do not pour any excess Art Roc down the sink. It may block it!

Further Things to Try

The versatility of this material and the finishes you could apply to it are endless, you can let your creativity flow. This material will also work equally well for free standing clocks. You can smooth the surface when working with it or leave it with a textured surface.

Clocks inspired by TEXTURES

... prickly, rough and course, hard, gravel and pumice, sand-baked, hardcore and concrete. Soft, smooth and shiny, silk-like fine and intricate, delicate, woven, soft. Wicker, wax and recessed, bristles, gorse and goose-grass, ...see it ...feel it, that's texture!

TEXTURED CLOCK

Clock Project 5

Starting Points	The textured clock is great for all those who love collecting bits and pieces of interest, such as shells, leaves, beads etc. This design can be made as shown or completely customised, by simply using the frame as a basis on which to stick everything. Your design can be as simple or as detailed as you like.
Design Features	I chose to use sand to give the clock an overall texture, which I then painted gold. I used string painted in a complementary colour to form the swirls.

TEXTURED CLOCK

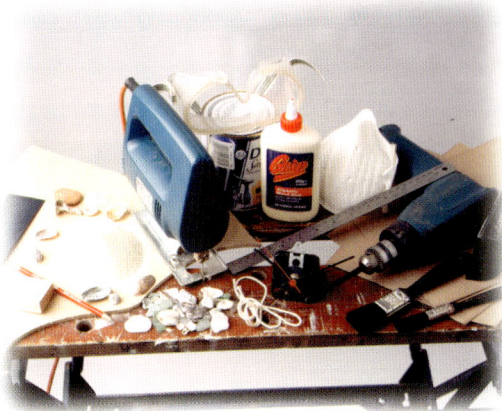

Equipment and materials used to make this clock.

Tools and Equipment
Drill and drill bit
Dust mask
Fine grade glasspaper
Jigsaw or coping saw
Paint brushes
Pencil & ruler
Safety goggles
Try-square

Materials
4mm MDF
All purpose adhesive
Clock mechanism
Masking tape (optional)
Paints or items of your choice
PVA glue
Sand
String

Instructions For Making

1. Mark out two identical squares on the MDF.
2. Cut out both squares, then mark and cut out another square on the inside of one of the pieces to create a frame.
3. Smooth each piece.
4. On the solid square, mark out the central point, by drawing diagonal lines from corner to corner. Drill a hole through the centre, the diameter of the spindle on your clock mechanism.
5. Glue both pieces together.
6. Once dry, cover the surface of the clock with glue and then cover in sand. To do this, lie the clock in the sand and then cover it over with more sand (see below).
7. When the glue is dry, brush off the excess sand.
8. Paint over the sand with brush or spray using your chosen colour.
9. Mark out the lines where the string will be glued onto the frame.
10. Paint the string.
11. Glue the string in place to form the pattern.
12. Fix the clock mechanism and hands into place.

Apply glue thickly to surface of MDF - mask off areas if required for straight edge.

Pour sand over entire piece of MDF and leave glue to dry.

Shake off excess sand and remove masking tape if the surface is to be one colour.

Paint using spray or brush. Remove any remaining masking tape.

Further Things to Try

Instead of applying sand to create a texture on the frame you may want to leave it smooth. You could develop a more detailed pattern with the string or you may want to stick things onto the frame, such as shells, glass beads, leaves, feathers or small pebbles, and then seal them with a coat of lacquer.

AIR DRYING CLAY CLOCK

Clock Project 6

Starting Points

Looking back at the different periods in history provides a wealth of material to inspire you. I chose the medieval period because I particularly like the designs of that era. The aspects I took as the inspiration for this clock were a jesters hat - which eventually became the shape of the clock, gold crowns – which I used to decorate the clock and the texture of deep purple, sumptuous velvet – which gave me the main colour.

Design Features

This clock has been made using air drying clay and, therefore, you do not need access to a kiln. This material is extremely versatile, and suitable for young children to use and achieve good results. This particular design has a relief pattern on the surface, and is open and hollow at the back.

AIR DRYING CLAY CLOCK

Equipment and materials used to make this clock.

Tools and Equipment
Rolling pin
Clay tools if needed (I used a craft knife and the tip of a ballpoint pen)
Paint brushes

Materials
Air drying clay
Clock mechanism
Cold water
Gold pen
Paints or finish of your choice

Instructions For Making

1. Very lightly sprinkle the work surface with cold water. The surface should be flat and smooth such as a laminate or wooden table top.
2. Roll out a lump of the clay using the rolling pin, to a thickness of approximately 4mm.
3. Cut out the shape you require using a craft knife. To make the design shown, I cut out the front shape, and then rolled a longer length for the long strip needed to go around the edge.
4. Join the two pieces together by laying the clock front down and then fitting the strip in place around the edge. Smooth the two pieces together using your finger and cold water. When they are joined at the back, carefully lift the entire clock upright, and smooth around the join at the front in the same way. (I didn't use a mould for this clock, but it would be advisable to use a mould covered in cling film to give the wet clay support whilst drying).
5. To make the crown details, roll out another piece of clay to a thickness of approximately 2mm. Cut the crowns out using a craft knife, and then wet the back of the crowns and place them onto the clock front, blending the edges with cold water. Make very small balls for the tops of the crowns and attach these in the same way.
6. Support the clock and leave it to dry. Just before it is fully hardened, push a pen through the centre of the clock face for the spindle on your clock mechanism. Thread the mechanism through the face to ensure it has clearance. Enlarge if required.
7. When it is completely dry and hard, paint straight onto the clay surface using watercolour paints. Then use a gold pen to add detail to the crowns.
8. Fit the clock mechanism into place.

Air drying clay

TIPS

Lightly wet your rolling pin with cold water to prevent it sticking to the clay.
Air drying clay can be white or pre-coloured.

Further Things to Try

The versatility of making the clay clock lies in the material itself. You can make any design you want with it. Different effects can also be achieved by the type of finish applied. I used watercolour paints, but acrylic paints would give the surface a sheen and the colours would be much stronger and brighter. You can also push materials into the surface of the clay to give it a texture or to leave an imprint, such as: metal mesh, sand, wire shapes, shells etc. You may even want to leave them in the clay.

FOLDED CLOCK WITH ETCHED FINISH

Clock Project 7

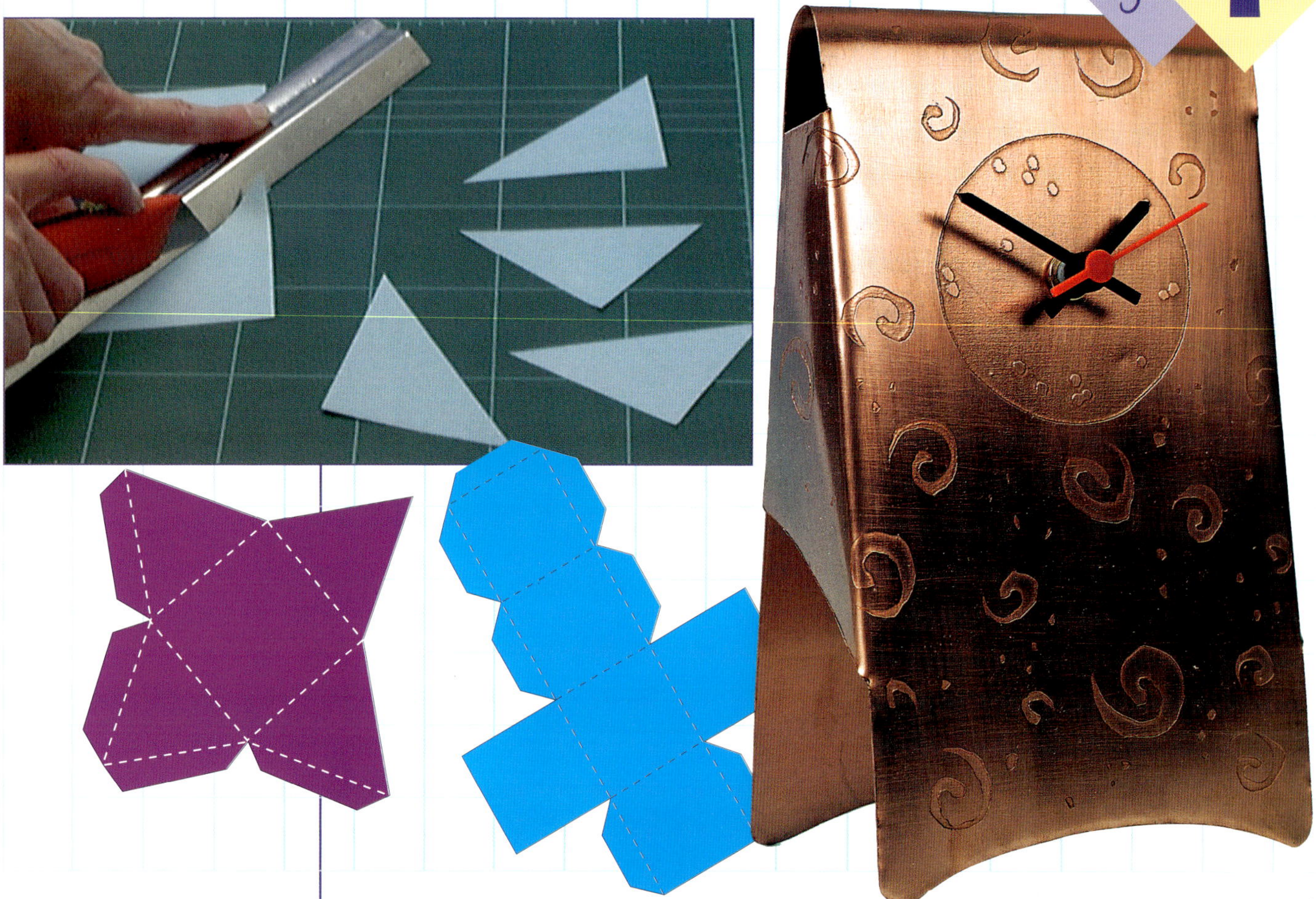

Starting Points	The idea for this clock developed from card modelling. It is made from one piece of material that has been folded into a free-standing shape.
Design Features	It is made from one piece of sheet copper that incorporates an etched technique which has been used for the surface decoration. The etching process enables you to create bold or intricate results. It is best to do a few test pieces before you etch your clock, so that you can experiment with the effect you want to achieve. (See page 50).
Tools and Equipment	Centre punch Dividers Drill and drill bit Fine grade steel wool Flat and half round file Hammer Leather mallet Paint brushes Piercing saw Rule Scriber Wet and dry paper
Materials	1mm Copper sheet. Clear lacquer Clock mechanism See details of etching on page 59 White spirit

Equipment and materials used to make this clock.

FOLDED CLOCK WITH ETCHED FINISH

Instructions For Making

1. Mark out the clock shape onto the copper, including the circle for the clock face, then centre punch the centre of the circle.
2. Cut out the design, and then mark your pattern onto it using a scriber.
3. File all the edges of the copper, then finish by rubbing them with a fine grade 'Wet and Dry' paper, such as grade 0.
4. Clean both sides of the metal using white spirit and steel wool, then rinse it in cold water.
5. To etch your pattern onto the clock, follow the instructions given on page 59 in this book.
6. Once you have etched your clock, drill a hole through the centre of the clock face the diameter size of the spindle of your clock mechanism, and then smooth the inside edge with a larger drill bit (de-burring).
7. Bend the clock into shape using a wooden jig. Copper is soft enough to bend by hand or can be gently tapped into shape using a leather mallet.
8. Clean up the copper by using white spirit and extra fine grade steel wool. Rub it over with a clean cloth.
9. Apply a coat of clear lacquer to protect the surface of the metal from discolouring.
10. Finally, fit the clock mechanism and hands into place.

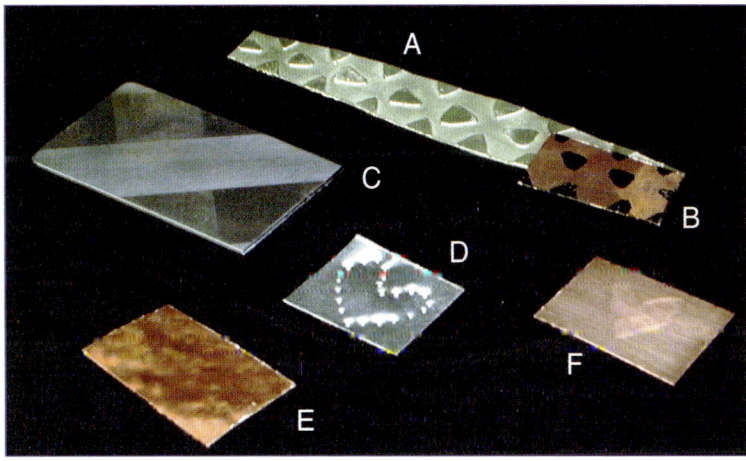

A. Raised pattern of etched copper. B. Stop-out varnish (shown black) applied to copper - area around the varnish is etched. C. Steel rubbed with wire wool with some areas covered using masking tape D. Pattern punched into metal. E. Planished copper. F. Etched pattern using a scriber.

TIPS

Make sure you cover all the edges with the stop out varnish so that they don't get eaten away during the etching process.

When cleaning the surface of the metal with steel wool, ensure you rub up and down in one direction only.

Further Things to Try

To develop this design further, you could incorporate some piercing work (cutting parts out of the surface of the metal), or enamel the etched areas. You can buy cold enamels which do not need to be heated in a kiln.

MOSAIC CLOCK

Clock Project **8**

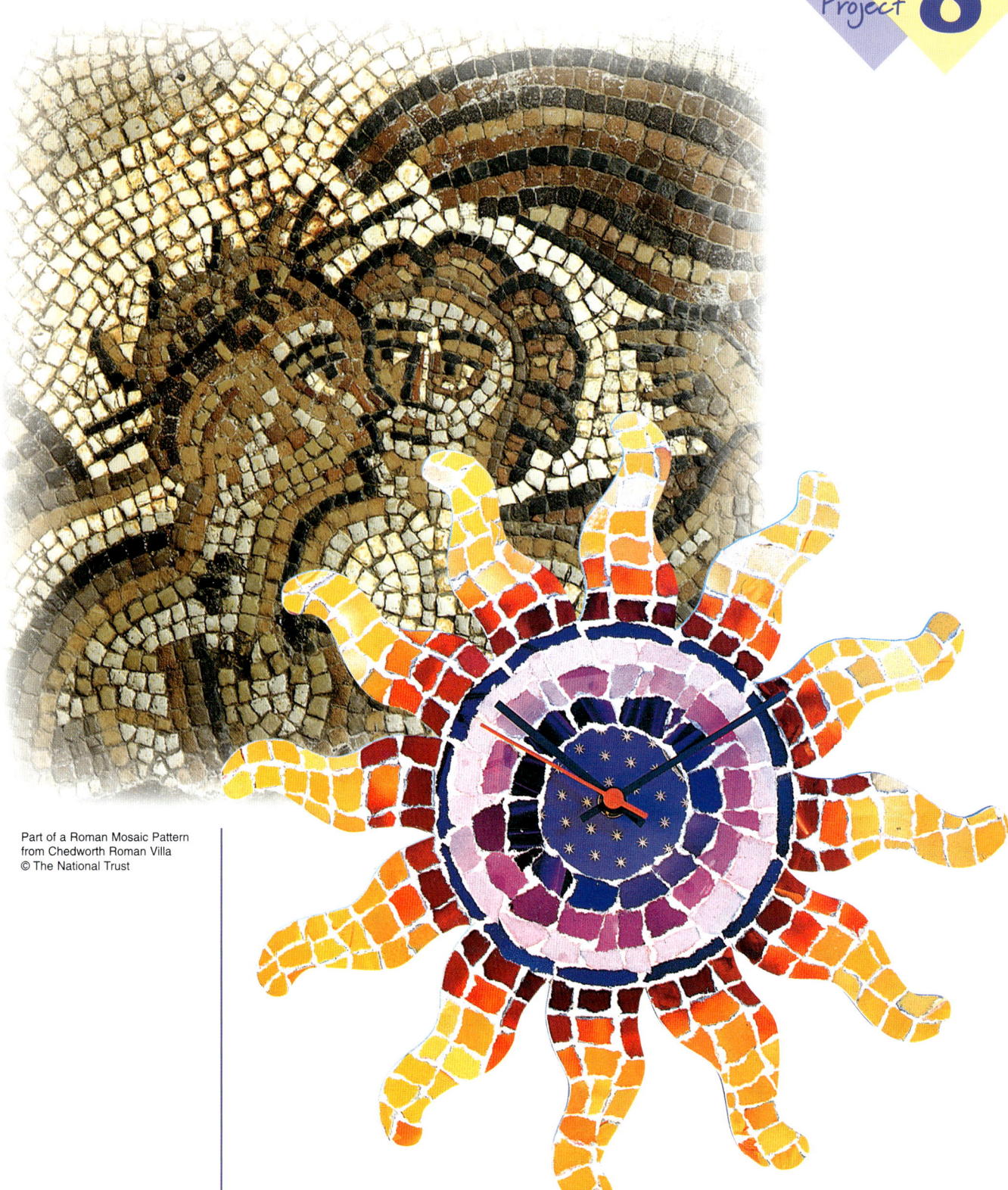

Part of a Roman Mosaic Pattern from Chedworth Roman Villa
© The National Trust

Starting Points | Mosaic is a lovely technique which always looks stunning. It reminds me of lost civilisations such as the Greeks and Romans and Mediterranean villas with their outdoor pools.... what better way to remind you of sunny holidays than a mosaic sun?

Design Features | This clock looks as if it is made using mosaic tiles, but it is actually made from thick card and paper torn out of old magazines. It is easy for all ages to make, and is another method which you could use to create any design you wish.

MOSAIC CLOCK

Tools and Equipment	Compass Cutting mat Craft knife Pencil Ruler
Materials	Clock mechanism Mounting board (Very thick card is also suitable) Old magazines Paper glue

Equipment and materials used to make this clock.

Instructions For Making

1. Mark out your chosen design onto the mount board (available from most artist suppliers or picture framers).
2. Cut out your design using the craft knife – adult supervision is advised for young children.
3. Outline your mosaic pattern in pencil
4. Create your mosaic design by tearing out the colours you want from the old magazines. It is a good idea to start in the middle of the design and work your way outwards. Try to keep your shapes similar, but they do not have to be the exact shape of your pencil outlines, they are a rough working guide.
5. Glue your mosaic pieces onto the mount board.
6. When it is dry, give it a coat of clear spray varnish to protect the surface, and to stop the edges of the paper from lifting up.
7. Carefully, make the hole for the mechanism in your design using a craft knife.
8. Fit the clock mechanism into place.

Further Things to Try

You can make this clock wall mounted or freestanding. To make it freestanding, you need another piece of mounting board cut into a rectangle to make the support (the same as you find on picture frames). Lightly score the board to enable it to bend and glue it to the back of your clock. Should you wish to develop your ideas further, there are plenty of books available on the market which cover mosaic techniques.

...structures tall and flexible, glass with reflective symmetry, bridges to infinity, Escher's endless staircase. Arches smooth and Roman, gothic and elaborate. Stone carved with precision, granite hard and polished... a visual feast of history.

Clocks inspired by

Architecture

TRADITIONAL MANTLE CLOCK

Clock Project 9

Starting Points	I have always been interested in the shapes and forms to be found in buildings. Many of the ideas in this book have developed from this. Often modern buildings take their architectural inspiration from the past. The clock I have designed here came from the shapes found in architecture.
Design Features	This clock is made from solid hardwood, and can be finished in a number of ways to co-ordinate with any interior scheme. To give the clock an 'aged' look which will make an eye catching feature, I have used a distressed paint effect. This has been done by painting the wood with two contrasting colours of paint, and then sanding away the top colour to reveal the lower colour beneath.
Tools and Equipment	Chisel and mallet (or mortise machine) Compass Drill and drill bits Fine grade glasspaper Jig-saw or tenon saw Paint brush Pencil Rule Try square
Materials	Adhesive Clock mechanism Hardwood and dowel Gold paint & 2 paint colours of your choice Stick-on numbers (I used Roman numerals)

Equipment and materials used to make this clock.

TRADITIONAL MANTLE CLOCK

Instructions For Making

1. Mark out each of the parts onto the timber.
2. Cut out all of the component parts.
3. On the main clock body, mark out the clock face.
4. Drill a hole through the centre of the clock face the diameter size of the spindle of your clock mechanism.
5. On the back of the main body, mark out a square, big enough to house your clock mechanism around the hole, and then rebate it down to a suitable thickness for the spindle length on your clock mechanism.
6. On the base, mark out where the main body will be attached, then do the same on the underside of the top triangular piece.
7. Mark out the drill holes for the dowel joints and columns on both the main body, base and the top, and drill each hole the diameter of your dowel and at least 10mm in depth. It is essential to ensure all the holes line up accurately.
8. Smooth all the parts down with fine grade glasspaper.
9. Glue and fix all the parts together.
10. Paint the clock with your first colour. If necessary apply two coats.
11. Paint the clock again with your second colour choice, again applying two coats if required.
12. Using a fine grade glasspaper, sand lightly over the clock taking away parts of the top paint colour, until the first colour begins to show through. You are aiming to achieve a worn/old look. Paint the clock face gold.
13. Mark out the number spacing on the clock face and put the numerals in place.
14. Fit the clock mechanism and hands into place.

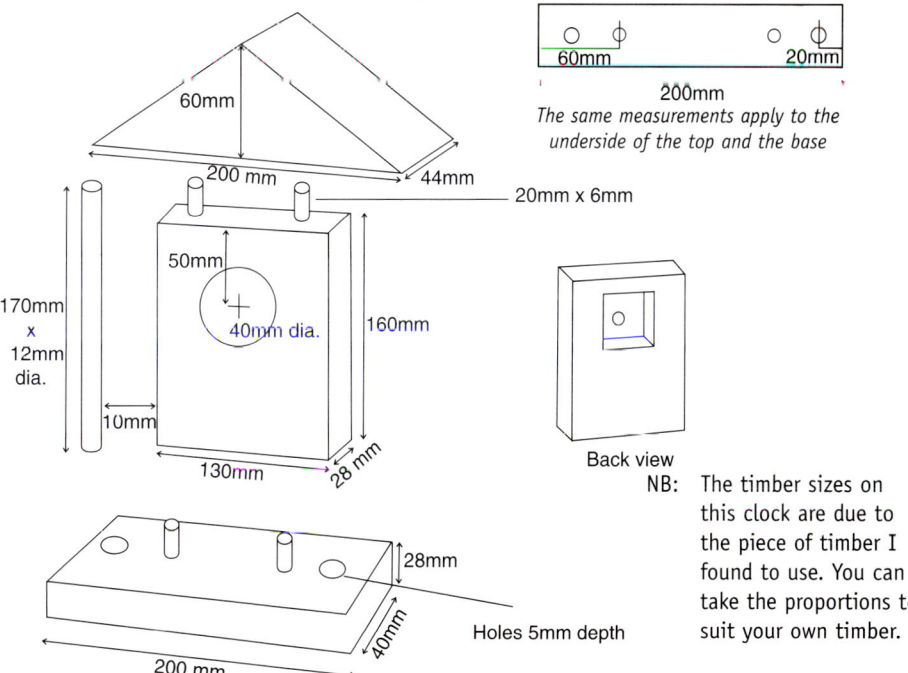

NB: The timber sizes on this clock are due to the piece of timber I found to use. You can take the proportions to suit your own timber.

TIPS — Most DIY stores sell sample pots of paint which are ideal for this sort of project.

Further Things to Try — You may want to keep the timber looking more natural, in which case you could finish it with a coat of beeswax or varnish for a more traditional look.

MANTLE/DESK CLOCK

Clock Project 10

Semi circular archways are a feature of Roman architecture

1940's oak cased mantle clock

Starting Points | This design has a mixture of both traditional and contemporary influences (the semi circular Roman arch and parallel flutes cut across the face of the timber) influences. The profile of the clock is a traditional shape for a mantle clock.

Design Features | This clock is made from a single piece of hardwood, which has just been given a coat of varnish to enhance the beauty of the timber. It is free standing and has very simple detailing that does not detract from the timber itself.

MANTLE/DESK CLOCK

Tools and Equipment

Compass
Drill and drill bit
Glasspaper
Jigsaw
Mortise machine or chisel and mallet
Pencil
Ruler
Router
Try-square

Materials

Beeswax or varnish
Clock mechanism
Hardwood timber

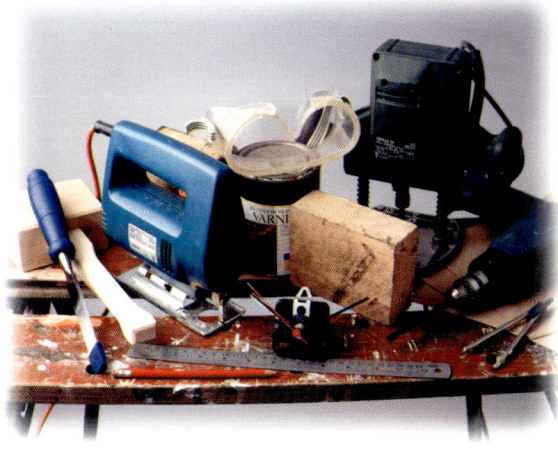

Equipment and materials used to make this clock.

Instructions For Making

1. Mark out the shape onto the timber.
2. Cut out the shape.
3. Mark out the lines on the front of the clock face to be rebated.
4. Using a router to rebate each line.
5. Mark out and drill the hole for the clock mechanism.
6. On the back of the clock, mark out a square around the hole a few millimetres larger than your clock mechanism.
7. Rebate this square using a mortise machine or by hand using a chisel or mallet, leaving an appropriate thickness suitable for the spindle length on your clock face.
8. Smooth the entire clock using glasspaper.
9. Finish the surface of the timber using beeswax or a varnish.
10. Fit the clock mechanism into place.

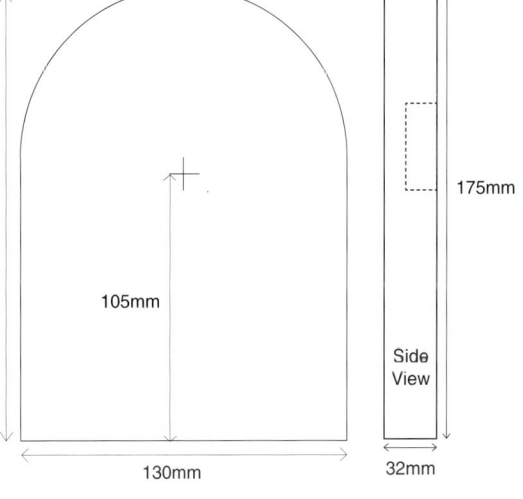

Further Things to Try

Use a router to make any pattern of your choice on the front of the timber. Highlight the detailing further by painting the parts that have been routed.

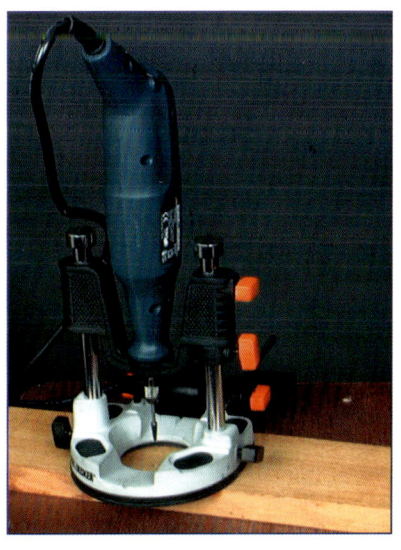

Router in use

ARCHED METAL CLOCK

Clock Project **11**

Starting Points — I took the bold shape of the arch as my starting point from which this design evolved. You can find a design in many everyday objects. Take a shape from a fabric design or a picture to co-ordinate your clock with your room. Just keep it simple and apply detail using colour and texture.

Design Features — I gave the clock an aged appearance by using a verdigris kit (readily available from DIY stores) which is a quick and easy method of making copper look old and weathered.

ARCHED METAL CLOCK

Tools and Equipment

Bench shears, Tinsnips or Junior hacksaw
Centre punch
Compass
Drill and drill bits
Flat and round files
Hammer
Masking tape
Pencil
Protractor
Rule
Try-square
Vice
Fine grade Wet and Dry Paper

Equipment and materials used to make this clock.

Materials

1mm thick Copper sheet
Clock mechanism
Verdigris kit or Patina materials (see page 58).

Instructions For Making

1. Cover both sides of the copper with masking tape to protect the surface from being scratched. Mark the design onto the metal, including the drill holes for the clock face.
2. Centre punch all the drill holes.
3. Cut the design out. You can use bench shears, tinsnips or a junior hacksaw.
4. Drill a hole through the centre of the clock face area the diameter size of the spindle on your clock mechanism. Drill the other twelve holes that represent the numbers, using a 4mm drill bit.
5. File all the edges of the metal, and the inside edges of the holes. Finish the edges by rubbing them with damp Wet and Dry paper, to get a very smooth finish.
6. Bend the base back to enable the clock to stand. (I placed the end of the clock into a vice to the depth I required and then bent the main part over).
7. Remove the masking tape and apply the finish using a bought verdigris kit or using the patina method following the corresponding instructions.
8. Fit the clock mechanism and hands into place.

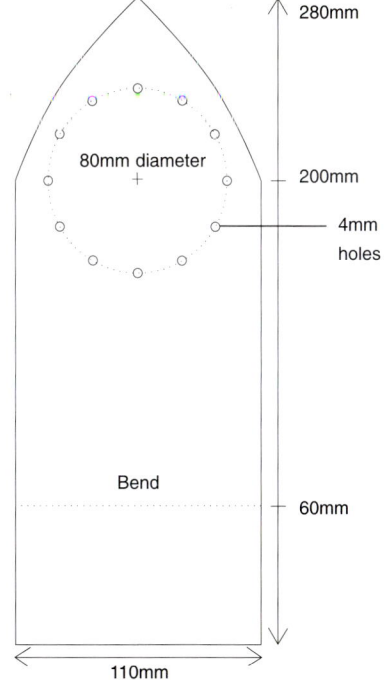

Further Things to Try

Alternatively, you may simply want to polish and seal the copper rather than giving it an aged effect, as it is a very attractive metal in itself. Another option is to put a different type of metal behind the holes to make them more of a feature. You could cut out a brass circle to fit behind the holes so that the brass colour shows through from the front. You can simply attach this to the back of the copper using adhesive, taking care not to let any show through the holes.

CUT & BEND CLOCK

Clock Project **12**

Starting Points

The principle behind this clock is in its name. It is made from one piece of material which has been cut and bent to form a free standing clock. The idea for the actual shape of the top and the detailing came from a Turkish building I saw in a travel brochure. This just goes to show that your initial ideas can come from anywhere!

Design Features

This design can be made in metal or acrylic depending on the finished look you want to achieve. It is also possible to make this design out of thick card like mounting board. I added extra detail by drilling small holes into the metal and punching, which is another technique you could use. Punch the metal from behind to create raised bumps on the front, using a centre punch and a hammer.

CUT & BEND CLOCK

Tools and Equipment

Depending on your choice of material
Coping saw or piercing saw
Drill and drill bit/s
Flat and round files
Masking tape
Pencil/pen
Rule
Wet and Dry paper
If using acrylic: Strip heater
If using metal:
Centre punch
Hammer
Lacquer
Paint Brush

Equipment and materials used to make this clock

Materials

A 1mm thick sheet of metal such as copper or 4mm acrylic sheet
Clock mechanism

Instructions For Making

1. If you are using metal, cover the surface with masking tape to protect it from getting scratched. If you are using acrylic leave the protective plastic cover on until you come to bend it. Mark the design out onto your chosen material, including the hole for the centre of the clock face (this will need to be centre punched if you are using metal).
2. Cut the design out.
3. Drill a hole through the centre of the clock face the diameter size of the spindle on your clock mechanism, and then smooth the inside edge.
4. File all the edges, and finish by rubbing them with fine Wet and Dry paper (for both metal and acrylic).
5. Add the detailing if desired – drill holes/punching.
6. Bend the legs back to the angle you want, using a jig to get each angle the same. If you are using acrylic, heat the area to be bent using a strip heater.
7. When you have bent the legs back, gently smooth the newly exposed edges.
8. If you have used metal, clean and polish the surface, and then apply a coat of clear lacquer to protect it from discolouring.
9. Fit the clock mechanism and the hands into place.

Further Things to Try

You can easily customise this design to make it your own, by changing the shape of the top, for example you could have it round or arched. You can add detail – if you use metal you can punch a pattern into it, you can add some piercing work into the design, give the clock a patina, or spray paint it. Similarly, if you use acrylic, you can also incorporate some piercing work into the design, or you could use a drill to make a pattern in it.

ACRYLIC CLOCK

Clock Project 13

Starting Points	This is a very simple design made from one piece of acrylic. The clock face and detailing have been kept in keeping with the simplicity of the design adding subtle interest. Acrylic is an easy material to work with and is available in a wide range of colours to suit any colour scheme.
Design Features	The idea for this clock came from modelling the shapes found in architecture with card, cutting out shapes and bending them. This is a very easy and inexpensive way of generating design ideas. You can use this technique to model free standing or wall mounted clocks. Because it is so easy to work with, you can experiment with different sizes and shapes against the wall or table to see how it looks. You can then also use your model as a template to help you mark out your material.

ACRYLIC CLOCK

Equipment and materials used to make this clock

Tools and Equipment
Ballpoint pen
Coping saw
Drill and 10, 8, 6 and 4 mm drill bits
Flat and round files
Rule
Sanding block
Strip heater
Try-square
Wet and Dry paper (medium & fine)
Wooden jig for 90° corners

Materials
4mm Acrylic sheet in a colour of your choice
Clock Mechanism

Instructions For Making

1. Leave the protective film on the acrylic to protect the surface from getting scratched. With a ballpoint pen, mark out the clock shape on the acrylic, including the drill holes and the centre for the clock face.

2. Cut the clock out.

3. Drill a hole the diameter of the spindle on your clock mechanism through the centre of the clock face. Then drill the 8, 6 and 4mm holes which form the decorative pattern.

4. File the edges of the acrylic until all the saw marks are removed and the edges are completely flat. Smooth the inside edges of the drill holes.

5. To achieve a good finish rub the edges with damp medium grade Wet and Dry paper, taking care not to catch the other surfaces of the acrylic (wrap the Wet & Dry paper around a wooden block). Finish with a fine grade for a really smooth edge.

6. Remove the protective plastic covering.

7. Heat the areas of the acrylic to be bent one at a time over a strip heater, and form them around a wooden jig to 90°. Hold each one in place until it is cool. (Remember the area that has been heated will be extremely hot).

8. Fit the clock mechanism into place.

Bending acrylic by using a C.R. Clarke Strip Heater

Further Things to Try

To increase the complexity of the design, you could inlay coloured acrylic rod into the decorative holes. You will need to ensure the ends of the rod are completely flat and flush with the surface of the acrylic clock. They can be secured in place using adhesive. You could also apply some surface decoration onto the acrylic by masking out a pattern using masking tape. Then rub the uncovered surface with steel wool in one direction to take the shine off the surface, thus creating very subtle detailing.

COPPER CLOCK

Clock Project 14

Starting Points
The material itself provided the idea for the design of this clock. Copper ages when exposed to the atmosphere forming copper oxide (verdigris) if it is not protected, and eventually develops a patina. I decided to contrast the new and the old in this design.

Design Features
The clock is made from one single piece of copper, so you don't need any special metalwork skills to make this clock. It has a verdigris edging to contrast with the warm tones of the metal, complemented by classical Roman numerals.

Tools and Equipment
Bench shears/tinsnips or a piercing saw.
Centre punch
Dividers or a compass and masking tape
Drill and drill bit
Flat file
Hammer
Pencil
Wet and Dry paper

Equipment and materials used to make this clock.

Materials
0.5 to 1mm thick copper sheet
Clock mechanism
Lacquer
Verdigris Kit

COPPER CLOCK

Instructions For Making

1. On the copper sheet, mark out the size of the circle you want your clock to be. You can use dividers to do this or you can put masking tape around the area you want to mark, and then use an ordinary compass and pencil to draw the circle onto the tape. Then centre punch the centre of the circle where the compass point has been.
2. Cut out the circle.
3. Drill a hole through the centre of the circle the diameter size of the spindle on your clock mechanism.
4. File all the edges of the copper.
5. Finish the edges by rubbing them with damp fine Wet and Dry paper, taking care not to scratch the front and back surfaces of the copper.
6. Clean the copper using white spirit and fine grade steel wool, taking care to rub in one direction only.
7. Mark out the border around the edge of the clock using a pencil and a compass or template.
8. Apply the verdigris to the border following the manufacturers instructions.
9. Give the entire clock a coat of lacquer and leave to dry.
10. Attach the self adhesive Roman numerals to the clock.
11. Fit the clock mechanism into place.

Using a verdigris paint effect on copper to achieve an 'aged' look.

 **TIPS**

To help you get a sharp edge between the two finishes you can mask off the centre to apply the verdigris around the edge. Wait until it is dry before removing the tape.

Further Things to Try

This design would look equally as striking as a square or rectangular shape. As an alternative to the verdigris border you could use the verdigris in the middle of the clock, leaving a copper border.

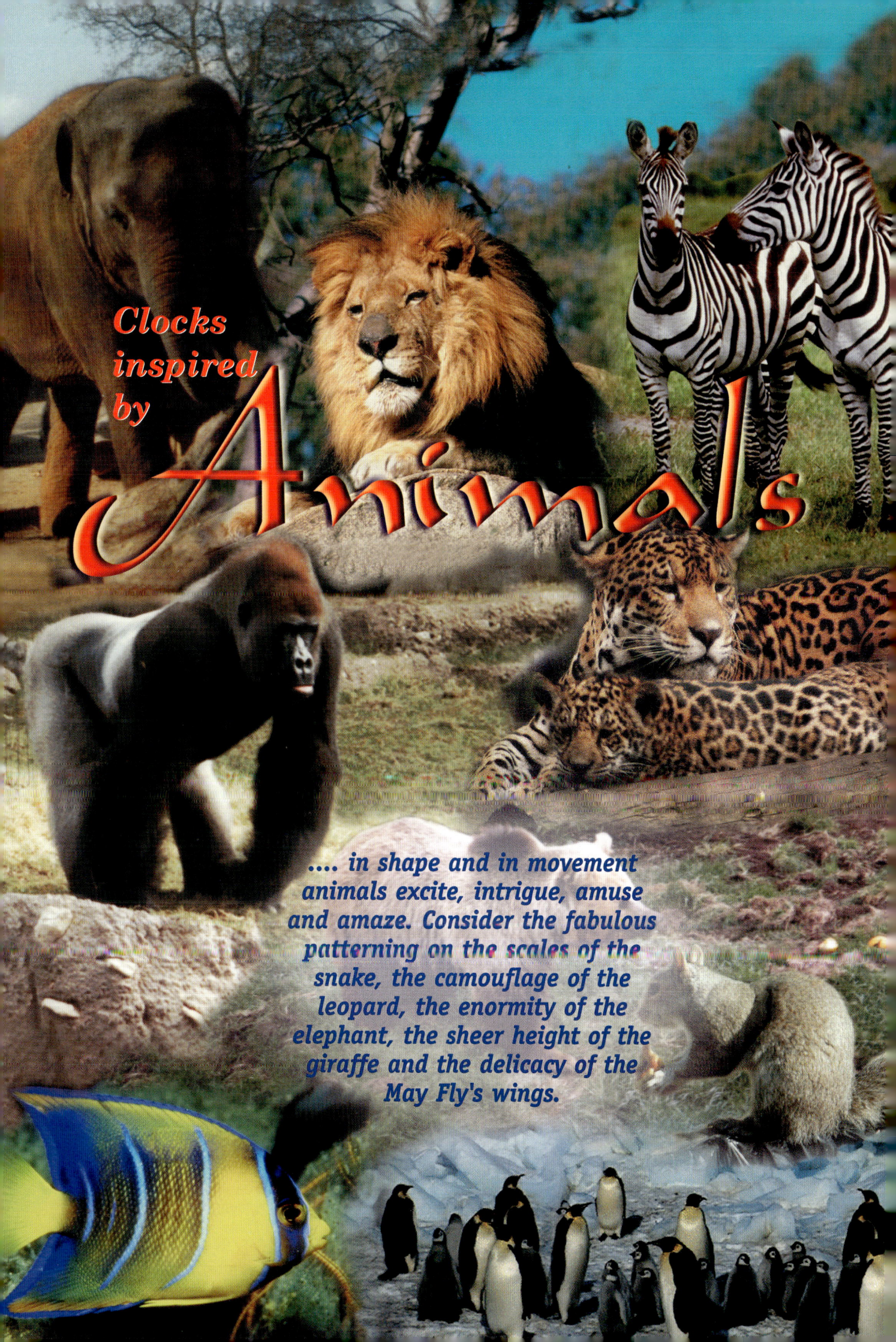

Clocks inspired by *Animals*

.... in shape and in movement animals excite, intrigue, amuse and amaze. Consider the fabulous patterning on the scales of the snake, the camouflage of the leopard, the enormity of the elephant, the sheer height of the giraffe and the delicacy of the May Fly's wings.

FUN PENDULUM CLOCKS

Clock Project **15**

Starting Points   The theme for these clock designs was fun! I chose animals as my focus, because they look fabulous, they are colourful, full of character and fascinating. I have studied animals moving and incorporated movement into my animal clock designs. My approach here is to over-emphasise certain features, much in the same way a cartoonist might work.

Design Features   These clocks are made in five easy steps! They are fun for all ages to make and enjoy. The three designs that you see here are all made using mounting board and paint. Children may want to make their favourite animal or character, or you can make a more humorous design for teenagers and adults. You can create personal designs based on interests and hobbies, which make these clocks great for gifts too.

FUN PENDULUM CLOCKS

Equipment and materials used to make this clock.

Tools and Equipment
- Cutting mat
- Craft knife
- Paint brush/es
- Pencil

Materials
- Clock mechanism
- Mounting board
- Paints
- Adhesive

Instructions For Making

Before you start to make your design, make a paper template to check that you have the overlap on the moving component correctly aligned. Try your clock mechanism with the templates to ensure the mechanism is hidden behind the main part of the clock, and that the moving part has enough overlap for movement and to attach to the pendulum. You need to ensure that no part will be seen that you don't want to be, when the pendulum swings.

1. Mark out your design in reverse onto the back of the mounting board.
2. Cut out the two parts for your design.
3. Paint the details onto your clock.
4. Glue the moving part onto the pendulum.
5. Fit the clock mechanism into place.

 **TIPS** Use a cutting mat or a piece of thick card underneath your work when you are cutting with a craft knife and always take great care when using a craft knife.

Further Things to Try

Fun pendulums can be based on anything. Children will love their favourite animal, or children's character. For adults you could base them on a favourite hobby or interest – such as fishing or golfing.

45

WHALE CLOCK

Clock Project 16

Starting Points | The sea and sea-life is a particularly favourite theme of mine. There are so many things from which to choose. I opted to make a whale, but you could select anything: a lighthouse, seahorse, beach hut, crab, octopus, boat, starfish etc. These clocks are of interest because of the mystery of the sea, they are fun, bright and amusing.

Design Features | This clock is very simple to make, and using this method you can make any design you wish. The shape is cut from one piece of wood, and the details are simply painted onto the surface of the clock face. What you must do is a select a shape with a strong profile which will be easily recognised.

WHALE CLOCK

Tools and Equipment
Drill and drill bit
Fine grade glasspaper
Jigsaw or coping saw
Paint brushes
Pencil
Rule

Materials
6mm MDF or plywood
Clock mechanism
White undercoat
Paints of your choice
Clear lacquer (optional)

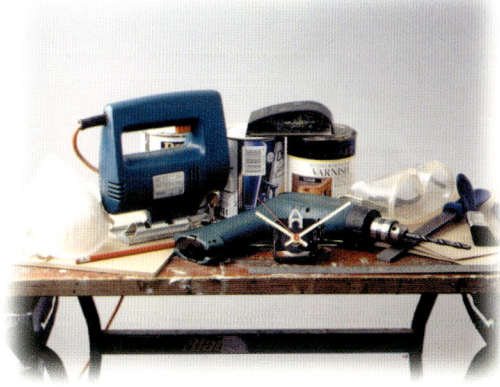
Equipment and materials used to make this clock.

Instructions For Making

1. Mark out the shape of the whale (or your own design) onto the wood.
2. Cut out the design.
3. Smooth the edge using a fine grade glasspaper. Do NOT smooth the front and back surfaces of MDF, otherwise the paint will not take to them as well.
4. Give the entire clock a coat of white undercoat, and leave to dry.
5. Mark out the details onto the clock.
6. Drill a hole through the centre of the clock face area the diameter size of the spindle on your clock mechanism.
7. Paint the details onto the clock using one colour at a time and leave it to dry before starting the next one to avoid smudges and runs (I used acrylic paint).
8. Apply a thin coat of clear spray varnish to protect the painted surface.
9. Fit the clock mechanism and hands into place.

Slippery Snake Clock is made using the same method as the Whale Clock and painted with gouache

TIPS
To gain a textured finish apply a thick coat of paint and almost pat it on.

Further Things to Try
Why not get your inspiration from a trip to the zoo or by looking at old vehicles?

CROCODILE CLOCK

Clock Project **17**

Starting Points	Animals are always popular, particularly with children. This clock is bright and good fun, and has been designed especially with children in mind. My inspiration for this clock came following a trip to the zoo. I chose a crocodile for my clock but I could just as easily have made the entire zoo! You could also take a child's favourite cartoon/book character and make it in the same way.
Design Features	I have used Formafoam® but it can be made with thick card, neoprene or similar materials which are easy to work with and cut easily with scissors or a craft knife. Each part is cut out separately and then glued into position so children can make it effortlessly.

CROCODILE CLOCK

Equipment and materials used to make this clock

Tools and Equipment
Compass
Pencil
Rule
Scissors or craft knife and cutting mat

Materials
Adhesive
Clock Mechanism
Dark green, light green, white and black Formafoam or you could use thick card, mount board, neoprene or similar modelling materials.

Instructions For Making

1. Mark out each of the parts on the Formafoam.
2. Cut out each of the pieces. (Children must be supervised at all times whilst using scissors, craft knives or glue).
3. Attach the shapes for the crocodile's back onto the body using glue. Apply it a few millimetres away from the edge, so that when it is pressed down onto the body, the glue won't run over the edges.
4. Attach the eye and the pupil in the same way.
5. Apply glue to the front of the teeth on the overlap, and then attach them into position from behind the crocodiles mouth, so that only the teeth are seen from the front.
6. Make a hole through the body the diameter size of the spindle on your clock mechanism, and then make another one through the centre of the clock face.
7. Glue the clock face onto the body of the crocodile.
8. Fit the clock mechanism and hands into place. (You can use a washer around the hole in the Formafoam to give it extra strength to hold the mechanism in place).

Formafoam is easy to cut to shape

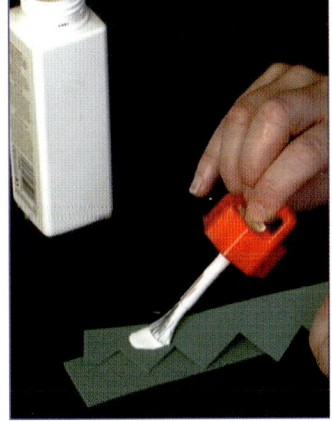

Applying glue to Formafoam

 **TIPS**

If you cannot draw well, find a picture of an animal or a character and enlarge it on the photocopier to the size you want then use this to draw around or trace.

Further Things to Try

This is an extremely easy way to make very attractive clocks, especially for small hands! You could make absolutely anything you wanted using it. Here are some more ideas: butterfly, rocket, hot air balloon, faces, other animals – lions, leopards, bears etc.

Novelty Clocks

... look for a strong profile shape, consider a nursery rhyme, think of a joke, encapsulate a saying, exaggerate and elaborate, design and distort, create an image, stylise and combine it into a time piece and you will have designed a novelty clock.

ZZZZzzz...CLOCK

Clock Project 18

Starting Points	There are lots of old sayings such as 'a stitch in time saves nine' and 'counting sheep to get to sleep', which is the one I chose for the design of this clock. I thought that this would go well in a child's bedroom. You can also use nursery rhymes or perhaps a poem to create a design.
Design Features	This clock is made from air drying clay. As no kiln is required and it can be easily painted, it is a suitable material for children to use.
Tools and Equipment	Craft knife or clay cutters and tools Pencil Rolling pin Paint brush/es
Materials	Air drying clay Clock mechanism Cold water Paints of your choice Self adhesive numbers

Equipment and materials used to make this clock.

ZZZZzzz... CLOCK

Instructions For Making

1. On a suitable surface, roll out the clay to a size big enough for your clock and to a thickness of approx. 5mm.
2. Cut out the main clock body. (I cut around a plate as a guide).
3. Make a hole in the centre of the circle the diameter size of the spindle on your clock mechanism. An easy way to find the middle if you have used a plate, is to draw around the plate onto paper, and then cut the circle out. Fold it in half and then in half again. Unfold the paper and where the two fold lines cross is the centre of the circle. Lay the paper over the clay and mark out the central point.
4. Roll out some more clay to a thickness of approximately 2mm and cut out twelve sheep. Place these around the edge of the clock and smooth around the edges of each one with water to join them to the clock.
5. Roll out a length of clay between the palms of your hand to make the fences. Attach in place in the same way and flatten them slightly.
6. Leave the clock to dry out thoroughly.
7. When it is dry paint the clock.
8. Attach the self adhesive numbers into place.
9. Fit the clock mechanism into place.

TIPS

You can use a normal table knife and other kitchen utensils to cut out and texture your clay shapes if you don't have clay tools.

Further Things to Try

You could create a wall clock for any room using this theme. Farm animal based designs make exceptionally great kitchen clocks. Jungle animals would make fun children's clocks, or maybe you know a cat or dog lover? A keen fisherman or golfer? the list is endless!

SAY CHEESE!

Clock Project **19**

Starting Points
If you are 'mad' about anything or you know someone who is, why not make a novelty clock. Something different will make a lovely and humorous gift.

Design Features
This clock is made in 'Art Maché' which can be used to create strong free-standing or wall mounted clocks in any design. You can achieve a smooth or textured finish which is easy to paint, or finish in a variety of ways.

Tools and Equipment
(Any craft tools you require)
Mould of your choice (I made my mould from an old cardboard box and masking tape)
Paint brush/es

Materials
Art Maché
Clock mechanism
Cold water
Paints of your choice (I used acrylic paints)

Equipment and materials used to make this clock.

Instructions For Making
1. Make your mould.
2. Make up the required amount of 'Art Maché' following the makers instructions on your pack.
3. Cover your mould with the maché, first cover the mould with tinfoil or cling film to aid the removal of the piece once it is dry. Remember to make a hole in the centre of the clock face area for the clock mechanism.
4. Leave it to dry thoroughly, and remove from the mould. If it is still damp underneath once you remove it from the mould, leave it to dry further.
5. Paint on your design.
6. Fit the clock mechanism into place.

Further Things to Try
This material would work well if you wanted to make a sculpture clock. To do this, use a fine metal mesh (such as the type you can buy for car body repairs) and shape it to the form you want. Ensure to have left enough room for the clock mechanism whether it is free-standing or wall mounted. Then cover your sculpture with the 'Art Maché' and finish as desired.

BIG NUMBERS CLOCK

Clock Project **20**

Starting Points

This clock has been designed with fun in mind! It is bold and bright, and will make a cheerful statement in any room, particularly a teenager's. Originally, the idea for this design came from a ball which had numbers all over it. When the ball rolled, the two colours appeared to be merging. I took all the numbers on a clock face and overlapped them into a circle. I then added a stripe around the edge and split the colours to reflect the movement of the ball.

Design Features

It is made of MDF and finished using paint. All the detail is in the painted design, which looks striking in two contrasting colours. I used a cream emulsion for the base colour, and then painted the numbers on with artists gouache. To further highlight this colour, I gently dabbed the red paint when it was still wet with a tissue to allow the base colour to come through in parts. I sealed this with two coats of spray varnish.

Tools and Equipment

Compass
Drill and drill bit
Dust mask
Fine grade glasspaper
Jigsaw or coping saw
Masking tape (optional)
Paint brushes
Pencil
Rule
Safety goggles

Materials

6mm MDF
Clock mechanism
Paint of your choice (I have used artists' gouache over a white emulsion undercoat and cream emulsion base colour).
Clear spray varnish

Equipment and materials used to make this clock.

BIG NUMBERS CLOCK

Instructions For Making

1. Using a compass, mark out a circle on the MDF and mark the centre of the circle with a small cross.
2. Cut the circle out of the MDF. NB Goggles and mask must be worn at all times when cutting or sanding MDF.
3. Drill a hole through the centre of the circle the diameter size of the spindle on your clock mechanism.
4. Finish the edges of the MDF by using fine grade glasspaper. Do not sand the front and back surfaces of the MDF otherwise the paint will not take to them.
5. Give the entire circle a coat of white undercoat. You made need to lightly sand the undercoat to remove any uneven parts.
6. Give the entire clock a coat of the cream emulsion. Then mark out the design.
7. Using your second colour paint the design onto the clock.
8. Seal the painted surfaces with a clear spray varnish and when dry apply another coat.
9. Fit the clock mechanism and hands into place.

 **TIPS**

If you don't want to paint the clock entirely free hand, mask off the areas you want to paint to get even lines. To do this, cover the areas with masking tape, cut out the parts you want to paint very lightly with a craft knife, then paint over it. Allow the paint to dry before you remove the masking tape.

Further Things to Try

To extend this design, you could cut the numbers out of thinner MDF and attach them to the clock face with adhesive to give the finished clock a raised face, and paint them in the same way.

Using MDF you could also make this aeroplane clock, pictured below. (See instructions for Whale Clock on page 46).

Details painted using watercolours and sealed with clear, spray varnish.

MAKING CLOCKS

HINTS & TIPS

MAKING CLOCKS HINTS AND TIPS

TIMBER

To mark out on timber use a pencil, ruler, try square and compass. To cut timber use a junior hacksaw for small straight cuts, a tenon saw for straight cuts, a coping or fret saw for curves and pierced work or an electric jigsaw for either type of cut. To shape and smooth timber use a plane on the surface, a file/disc sander on the edges and glasspaper all over. The exception to the rule is MDF, where you should only sand the edges. To finish timber use paint, wood stain, beeswax or varnish.
N.B. Safety: Always take care when using a cutting knife. When sawing MDF it is important that you wear safety goggles and mask.

METAL

To mark out metal, use marking blue and a scriber with an engineer's square, ruler, centre punch and dividers. Alternatively, cover the surface of the metal with masking tape and use a lead pencil and a compass instead of scriber and dividers. This also helps protect the metal from getting scratched as you work with it. To cut metal use a junior hacksaw for small straight cuts, a hacksaw or bench shears for straight cuts and a piercing saw for curves and pierced work. Straight and curved tinsnips are also very useful. To shape and smooth metal, use a file on the edges and then a suitable abrasive paper to finish. When drilling metal, it should be clamped down over a piece of timber for every hole. To finish metal, polish the surface, buff it, paint it, apply a coat of clear lacquer or a surface treatment to colour it. To bend thin metal sheet in a straight line, use a folding bar in a vice, or use the edge of a bench as a guide.

ACRYLIC

Acrylic comes with a clear protective film on both sides to protect it from getting scratched, which you should leave on where possible. Mark it out using a ballpoint pen. To cut acrylic use a junior hacksaw and a coping saw. Shape and smooth the edges of acrylic as you would metal. When drilling acrylic it should be clamped down over a piece of timber for every hole, and you should drill slowly. Grinding the 'land' away inside a twist drill bit helps to prevent it from 'scratching' as it cuts through thin plastic sheet. The edges can also be polished and buffed. To bend acrylic sheet, you should heat both sides of the acrylic over the line to be bent using a strip heater. When the acrylic has softened bend it to shape . If you want to bend it to a specific angle or you want two bends the same, you should use a wooden mould to bend it over or around. Remember the acrylic will be hot when heated and care should be taken when handling it, use leather gloves as a precaution.

COPPER VERDIGRIS PATINA FINISH

This patina is greenish-blue, and takes the form of a stippled incrustation over the surface of the copper. To make the solution you will need :

Ammonium carbonate 120gm	Ammonium chloride 40gm
Sodium chloride 40gm	Water 1 litre
Sawdust or wood shavings	Wood or plastic container big enough to hold your work.

Pour the solution over the sawdust in your container ensuring it is all moistened. Place the copper to be coloured into the moistened sawdust making sure that it is completely covered. Leave the piece in the solution for 20 – 25 hours. The sawdust should be kept moist throughout the treatment. When the patina has developed remove it from the sawdust and rinse it gently under cold water, and leave it to dry. If the patina is too heavy you can lightly brush it with a wire brush to remove some of the incrustation. To finish, you can apply a coat of wax or clear spray lacquer.

ALTERNATIVELY, you can achieve a similar effect using a verdigris kit available at most DIY stores. They use paints to create the effect rather than chemicals.

MAKING HINTS AND TIPS

ETCHING COPPER

You can achieve two results when etching copper.

1. You can etch a pattern into the copper so that it is recessed.
2. You can etch a pattern into the copper so that it is raised.

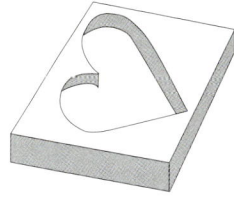
Recessed

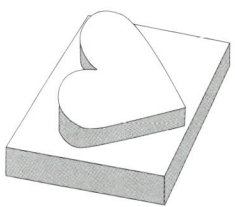
Raised

You will need;
Cold water
Glass or enamelled container/sink (big enough to lay your clock in)
Steel wool
Nitric acid
Stop out varnish
Tongs
Safety goggles, overalls and leather gloves

Before you begin to etch copper you must ensure the surface is completely cleaned. Use white spirit and/or extra fine grade steel wool to clean it. If you want to achieve a recessed design, you will need to paint the copper around the design leaving the design uncovered. If you want to achieve a raised effect, you will need to paint the entire surface apart from the raised pattern area required. For both effects paint out the edges and the back completely. Ensure there are no small gaps in the stop out varnish otherwise the copper will be eaten away. Once you have painted your design onto the copper using the stop out varnish and it is thoroughly dry, immerse it into the solution (1 part acid to 1 part water). Leave it, checking it regularly until the design is etched to the depth you require. Then remove it from the solution, **using tongs**, and rinse it under cold water. Remove the remaining stop out varnish using white spirit, and steel wool, then rinse it again with cold water. The effect can take anything from minutes to a couple of hours depending on the depth of etch you want. Keep checking it and remove your work when it has etched enough.

(**NB:** Always add the acid to the water – never the other way around. Just make enough solution to cover your work completely).

PAPER MAKING

This is a basic paper making technique.
To make paper to form around a mould you will need :
Waste papers/fibres preferably a suitable colour
Liquidizer
Water

The waste paper you use will determine the look of the paper you make, and its characteristics. For the handmade clock it is preferable to use old water-colour paper, which will make it stronger than if you use a thinner paper. To make your paper you first need to tear it into small pieces. Then put a handful of the paper pieces into the liquidizer which should be two thirds full of water, and blend the pieces into a pulp. For this particular clock you do not want the pulp to be too smooth, because you want to create a nice texture. To strengthen the paper, you can add size to the pulp, such as cold water starch, or PVA. You can also add watercolour paint to the pulp to colour it. You will have to repeat this pulp making process until you have enough to cover your mould. The next stage is to apply the pulp to the mould, using your hands. Empty the pulp into another container, squeezing the majority of the water out. Cover the entire mould with the pulp, leaving no gaps, pressing the pulp down and together. Finish as required. Leave it to dry thoroughly, and then gently lift your paper clock from the mould.

CLOCK FACE NUMERALS TEMPLATE

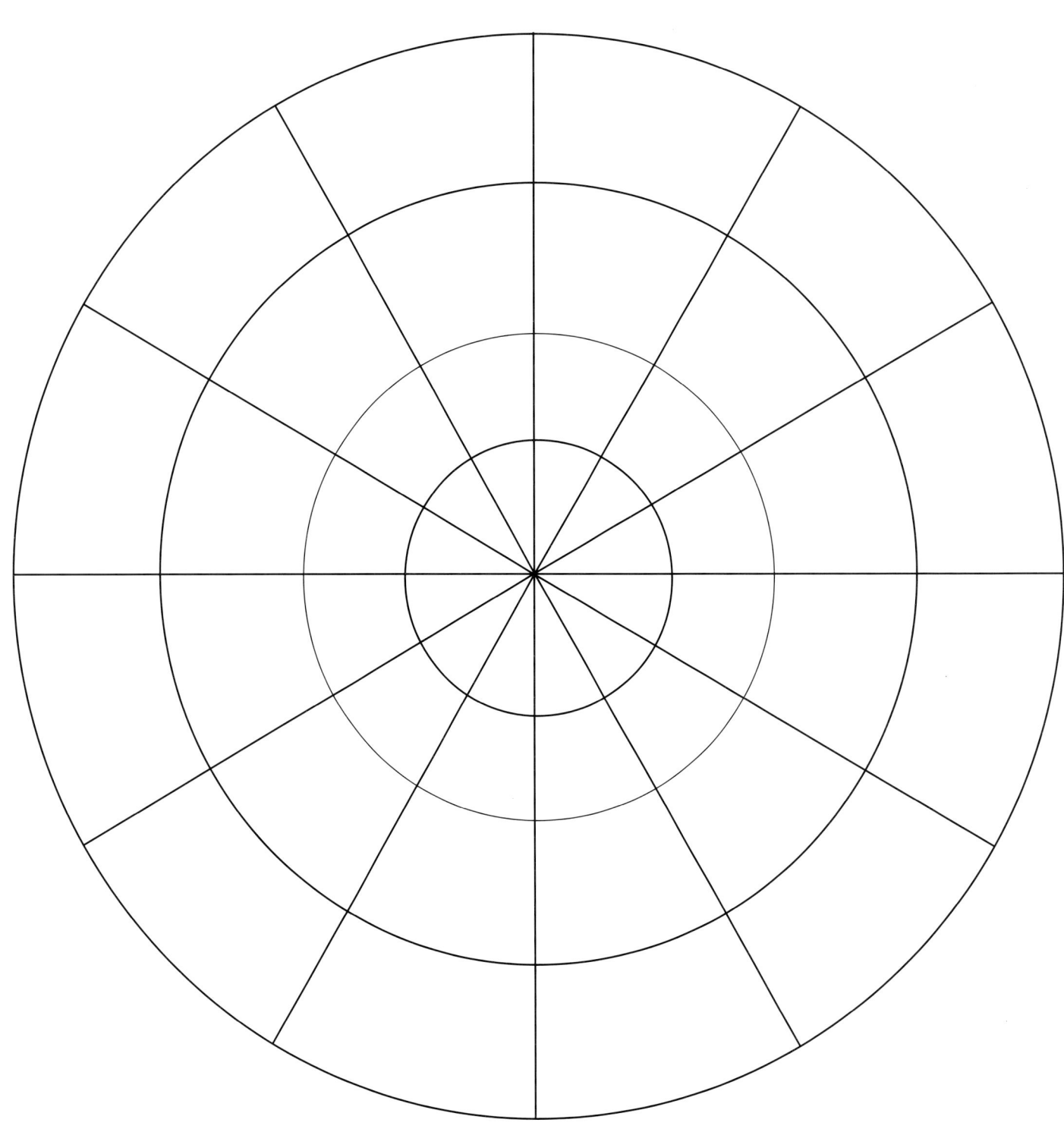